LET'S LOOK AT LANDFORMS

THAT'S A PLATEAU!

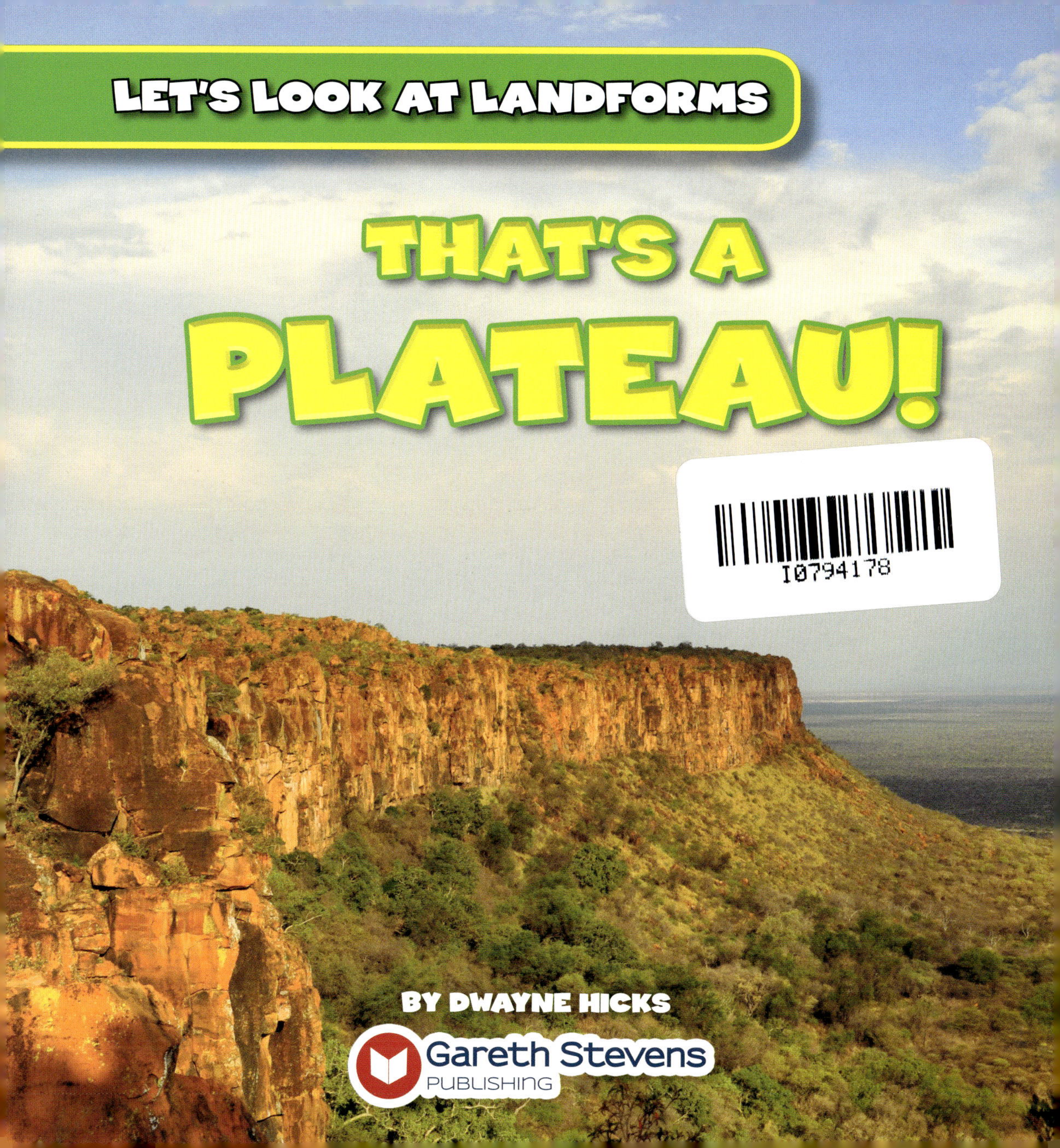

BY DWAYNE HICKS

Gareth Stevens PUBLISHING

Please visit our website, www.garethstevens.com. For a free color catalog of all our high-quality books, call toll free 1-800-542-2595 or fax 1-877-542-2596.

Library of Congress Cataloging-in-Publication Data
Names: Hicks, Dwayne, author.
Title: That's a plateau! / Dwayne Hicks.
Other titles: That is a plateau!
Description: New York : Gareth Stevens, [2022] | Series: Let's look at landforms | Includes index.
Identifiers: LCCN 2020032259 (print) | LCCN 2020032260 (ebook) | ISBN 9781538263754 (library binding) | ISBN 9781538263730 (paperback) | ISBN 9781538263747 (set) | ISBN 9781538263761 (ebook)
Subjects: LCSH: Plateaus–Juvenile literature.
Classification: LCC GB571 .H53 2022 (print) | LCC GB571 (ebook) | DDC 551.43/4–dc23
LC record available at https://lccn.loc.gov/2020032259
LC ebook record available at https://lccn.loc.gov/2020032260

Published in 2022 by
Gareth Stevens Publishing
29 E. 21st Street
New York, NY 10010

Designer: Tanya Dellaccio
Editor: Greg Roza

Photo credits: Cover brytta/E+/Getty Images; pp. 3–22 (background texture) MicroOne/Shutterstock.com; p. 5 Feargus Cooney/Lonely Planet Images/Getty Images Plus/Getty Images; p. 7 Photofusion/Unverisal Images Group/Getty Images; p. 9 NurPhoto/Getty Images; p. 11 4nadia/iStock/Getty Images Plus/Getty Images; p. 13 sunsinger/Shutterstock.com; p. 15 Elena_Suvorova/Shutterstock.com; p. 17 Philippe Bourseiller/The Image Bank/Getty Images Plus/Getty Images; p. 19 Elena Zhi/Shutterstock.com; p. 21 Jim Dyson/Getty Images News/Getty Images.

Printed in the United States of America

CPSIA compliance information: Batch #CSGS22: For further information contact Gareth Stevens, New York, New York at 1-800-542-2595.

CONTENTS

Boldface words appear in the glossary.

Tablelands

A plateau is a large, flat area of land. Plateaus are sometimes called tablelands. Some plateaus are covered with plains and farmland. Others are deserts. The pyramids near Cairo, Egypt, were built on the Giza Plateau.

High Plains

Plateaus are located at high **elevations**. Because of this, they're sometimes called high plains. Plateaus are often higher than areas around them. Some end at **steep** hills or **cliffs**. Some are bordered by mountains or oceans.

Pushing Up

Sometimes, **molten** rock from deep inside Earth comes close to Earth's **surface**. This creates a lot of **pressure**! The pressure pushes the land above it higher than surrounding areas. This creates a plateau. It can also create volcanoes!

Flowing Out

Volcanoes are breaks in Earth's surface that allow molten rock to pour out. This rock cools and hardens to form new rock. Some plateaus were made when molten rock flowed out of a volcano and covered a large area.

Piling Up

Earth's surface is made out of "plates" that fit together like puzzle pieces. Sometimes the plates run into each other. One plate can be forced under the other. One plate rises higher than the other. This creates mountains and plateaus.

Shaping Up

Erosion can create plateaus. Wind and water wear away mountains and leave the dirt and rock between them. This creates a plateau surrounded by mountains. Wind and water can cut plateaus into amazing shapes. Erosion can make smaller plateaus called mesas.

Cutting Up

Some plateaus have many crossing rivers. The rivers get deeper over many years. This results in flat-topped land separated by valleys and **canyons**. This type of plateau is called a dissected plateau. "Dissected" means "cut into parts."

The Tibetan Plateau

The Tibetan Plateau in Asia is the largest plateau in the world. It's almost four times as big as Texas! It's also the highest plateau in the world. Most of it is higher than 13,100 feet (4,000 m) above sea level!

La Paz, Bolivia

Plateaus cover about one-third of Earth's surface! People have built cities on plateaus all over the world. The Altiplano Plateau in South America is home to many cities. La Paz, Bolivia, is more than 10,500 feet (3,200 m) above sea level!

GLOSSARY

canyon: a deep valley with V-shaped sides that usually has a river at the bottom

cliff: a tall, steep area of land

elevation: the height something is above sea level

erosion: the wearing away of the earth by wind or water

molten: in a liquid form

pressure: the force of one thing pressing or squeezing another

steep: almost straight up and down

surface: the outer layer of something

FOR MORE INFORMATION

BOOKS

Capelle, Jenna. *Plateaus*. Mendota Heights, MN: North Star Editions, 2018.

Somervill, Barbara A. *Plains and Plateaus*. North Mankato, MN: Cherry Lake Publishing, 2014.

WEBSITES

Plateau

www.nationalgeographic.org/encyclopedia/plateau/

Read about plateaus on National Geographic's website.

What Is a Plateau?

eschooltoday.com/landforms/what-is-a-plateau-landform.html

Find out more about plateaus and other landforms here.

Publisher's note to educators and parents: Our editors have carefully reviewed these websites to ensure that they are suitable for students. Many websites change frequently, however, and we cannot guarantee that a site's future contents will continue to meet our high standards of quality and educational value. Be advised that students should be closely supervised whenever they access the internet.

INDEX